4th International Conference on Plant Synthetic Biology, Bioengineering & Biotechnology 2020

Online
30 October – 1 November 2020

ISBN: 978-1-7138-2774-0

Printed from e-media with permission by:

Curran Associates, Inc.
57 Morehouse Lane
Red Hook, NY 12571

AIChE
120 Wall Street, FL 23
New York, NY 10005-4020

Phone: (800) 242-4363
Fax: (203) 775-5177

www.aiche.org

Additional copies of this publication are available from:

Curran Associates, Inc.
57 Morehouse Lane
Red Hook, NY 12571 USA
Phone: 845-758-0400
Fax: 845-758-2633
Email: curran@proceedings.com
Web: www.proceedings.com

TABLE OF CONTENTS

Diversity, Structure, Function, Assembly and Engineering of Carboxysomes and Other Bacterial Microcompartments for Plant Synthetic Biology

Conference: International Conference on Plant Synthetic Biology and Bioengineering

Year: 2020

Proceeding: 4th International Conference on Plant Synthetic Biology, Bioengineering, and Biotechnology

Group: General Submissions

Session: Keynote 1

Time: Friday, October 30, 2020 - 10:00am-10:25am

Authors:

Kerfeld, C. A., Department of Energy Joint Genome Institute and University of California, Berkeley

Carboxysomes and other bacterial microcompartments are multienzyme-containing organelles composed entirely of protein. The carboxysome is a self-assembling metabolic module for CO_2fixation found in all cyanobacteria. These large (~100-500 nm) polyhedral bodies sequester Carbonic Anhydrase and RuBisCO within a protein shell, thereby concentrating substrates and reducing photorespiration. Because carboxysomes and other BMCs function to organize reactions that require special conditions for optimization, including the sequestration of substrates, (sensitive) cofactors, or toxic intermediates and the protection of oxygen-sensitive enzymes, they have received considerable attention as templates for synthetic nanoreactors or scaffolding in bioengineering. There are two central challenges to building bespoke protein-based nanoreactors, design and assembly of multi-enzyme cores and engineering of the shell proteins to serve as a selectively permeable barrier to metabolites and electrons, thereby providing the interface between the cytosol and the encapsulated reactions. Progress in our efforts to engineer carboxysomes and, more broadly to take advantage of protein-based compartmentalization in bioengineering, will be discussed.

Compartmentalizing Terpenoid Bioproduction in Plants with Engineered Plastid Pathways and Lipid Droplet Scaffolds

Conference: International Conference on Plant Synthetic Biology and Bioengineering

Year: 2020

Proceeding: 4th International Conference on Plant Synthetic Biology, Bioengineering, and Biotechnology

Group: General Submissions

Session: Oganelles and Synthetic Compartments

Time: Friday, October 30, 2020 - 10:50am-11:15am

Authors:

Bibik, J. D., Michigan State University
Banerjee, A., Michigan State University
Weraduwage, S. M., Michigan State University
Robertson, K., Michigan State University
Sharkey, T. D., Michigan State University
Hamberger, B., Michigan State University

Leveraging synthetic biology approaches, engineered plants offer a sustainable production platform for high-value chemicals and other bioproducts. Squalene, a C30 hydrocarbon, is a biofuel candidate and the precursor to high-value triterpenoids, a diverse class of natural products with applications in the health, cosmetic, and other biotechnological industries. In this work, two strategies have been developed to increase plant production yields of squalene and triterpenoids by hijacking existing cell compartments or building novel compartments to sequester products within cells. Natively, squalene biosynthesis occurs in the cytosol through farnesyl diphosphate synthase (FDPS) and squalene synthase (SQS). The first strategy re-localizes this pathway to plastids, natural intracellular compartments, where biosynthesis can occur separate from native competing enzymes. In plastids, squalene yields were optimized through screening of diverse orthologs and engineered variants of key steps in the pathway. The second strategy re-engineers cytosolic lipid droplets as synthetic storage organelles with biosynthetic enzymes anchored to the surface, synthesizing and storing products in the same location. Fusing FDPS and SQS to the algal Lipid Droplet Surface Protein (LDSP) enables enzyme scaffolding on the surface of lipid droplets, which sequester the hydrophobic products. Preliminary work also demonstrates targeting the SQS-LDSP-FDPS fusion protein to plastids mediates negative effects on photosynthesis, possibly through novel scaffolding and further squalene sequestration. These strategies effectively increased squalene yields using transient expression in *N. benthamiana* and are being implemented in stable poplar transformants, a target production crop. Effective compartment engineering improves plant production of squalene, while also providing platforms to expand towards higher-value triterpenoids and other bioproducts.

The Pyrenoid: A Liquid-Liquid Phase Separated CO2 Fixing Organelle

Conference: International Conference on Plant Synthetic Biology and Bioengineering

Year: 2020

Proceeding: 4th International Conference on Plant Synthetic Biology, Bioengineering, and Biotechnology

Group: General Submissions

Session: Oganelles and Synthetic Compartments

Time: Friday, October 30, 2020 - 11:15am-11:40am

Authors:

Mackinder, L., University of York

Approximately one-third of global carbon-fixation occurs in an overlooked algal organelle called the pyrenoid. The pyrenoid contains the CO_2-fixing enzyme Rubisco, and enhances carbon-fixation by supplying Rubisco with a high concentration of CO_2. Since the discovery of the pyrenoid over 130 years ago, the molecular structure and biogenesis of this ecologically fundamental organelle have remained enigmatic. To improve our understanding of the pyrenoid, we have developed high-throughput fluorescent protein-tagging pipelines to determine the localizations of over 250 candidate proteins in the green alga *Chlamydomonas reinhardtii*. Combining localization data with protein-protein interaction data has allowed us to generate a spatial interactome of the pyrenoid. This has led to the discovery that the pyrenoid is a highly dynamic non-membrane bound organelle that has liquid-like properties and can undergo rapid liquid-liquid phase separation. This Rubisco condensation is mediated by a low complexity repeat protein that links Rubisco molecules and enables phase separation. These new insights are guiding the engineering of the pyrenoid into higher plants with a goal to enhance crop carbon fixation efficiency.

Towards Engineering a Carboxysome Based CCM into C3 Crops

Conference: International Conference on Plant Synthetic Biology and Bioengineering

Year: 2020

Proceeding: 4th International Conference on Plant Synthetic Biology, Bioengineering, and Biotechnology

Group: General Submissions

Session: Oganelles and Synthetic Compartments

Time: Friday, October 30, 2020 - 11:40am-12:05pm

Authors:

Parry, M., Lancaster University

Improving the efficiency of photosynthesis is widely recognized as a promising strategy for increasing the yield of biomass and food crops. Most of the research aimed at improving photosynthesis is focused on overcoming the limitations of the primary CO_2 fixing enzyme Rubisco. In cyanobacteria, bacterial microcompartments, called carboxysomes, improve the efficiency of photosynthesis by acting as a carbon concentrating mechanism (CCM) to concentrate CO_2 near Rubisco. Cyanobacteria also make use of Rubiscos that are faster than those of crop plants, although they have lower specificity toward CO_2 than the land plant enzyme, which increases resource use efficiency. Replacement of land plant Rubisco by faster bacterial variants with lower CO_2 specificity will improve photosynthesis and resource use efficiency if installed in a microcompartment capable of concentrating CO_2. Here we will report our recent progress on introducing components of the cyanobacterial CCM into plants.

In silico Model for Mining the Cis-Regulatory Determinants of Tissue-Specific Gene Expression

Conference: International Conference on Plant Synthetic Biology and Bioengineering

Year: 2020

Proceeding: 4th International Conference on Plant Synthetic Biology, Bioengineering, and Biotechnology

Group: General Submissions

Session: Genome Scale Engineering

Time: Friday, October 30, 2020 - 12:00pm-12:25pm

Authors:

Fraser, V. N., Oregon State University
Ansariola, M., Oregon State University
Filichkin, S., Oregon State University
Ivanchenko, M. G., Oregon State University
Bright, Z. A., Oregon State University
Gould, R. A., Oregon State University
Ozguc, O. R., Oregon State University
O'Neil, S., Oregon State University
Megraw, M., Oregon State University

Gene expression across tissues is regulated by a combination of determinants, including the binding of transcription factors (TFs), along with other aspects of cellular state. Recent studies emphasize the importance of both genetic and epigenetic states â€" TF binding sites and binding site chromatin accessibility have emerged as potentially causal determinants of tissue specificity. To investigate the relative contributions of these determinants, we constructed three genome-scale datasets for both root and shoot tissues of the same *Arabidopsis thaliana* plants: TSS-seq data to identify Transcription Start Sites, OC-seq data to identify regions of Open Chromatin, and RNA-seq data to assess gene expression levels. For genes that are differentially expressed between root and shoot, we constructed a machine learning model to predict tissue of expression from chromatin accessibility and TF binding information upstream of TSS locations. The resulting model was highly accurate (over 90% auROC and auPRC), and our analysis of model contributions (feature weights) strongly suggests that patterns of TF binding sites within ~500 nt TSS-proximal regions are predominant explainers of tissue of expression in most cases. Thus, in plants, cis-regulatory control of tissue-specific gene expression appears to be primarily determined by TSS-proximal sequences, and rarely by distal enhancer-like accessible chromatin regions. This study highlights the exciting future possibility of a native TF site-based design process for the tissue-specific targeting of plant gene promoters.

Reconstructing Regulation: Can Biological Complexity be Reverse Engineered?

Conference: International Conference on Plant Synthetic Biology and Bioengineering

Year: 2020

Proceeding: 4th International Conference on Plant Synthetic Biology, Bioengineering, and Biotechnology

Group: General Submissions

Session: Genome Scale Engineering

Time: Friday, October 30, 2020 - 12:25pm-12:50pm

Authors:

Patron, N., Earlham Institute

Plants provide the potential for rapid production of complex molecules from water and light but, until recently, we lacked the tools and data necessary for complex engineering of plant systems. Until recently, a paucity of data for cis-regulatory elements in plants meant that it was challenging to determine how individual sequence elements contributed to transcriptional function. We have identified functional elements and established a quantitative experimental system to investigate transcriptional function, investigating how identity, density and position contribute to regulatory function. We then identified permissive architectures for minimal synthetic plant promoters enabling the computational design of a suite of synthetic promoters of different strengths. These have been used to regulate the relative expression of output genes in simple genetic devices. We are applying these methods to engineer plants as photosynthetic biomanufacturing platforms and to improve agricultural and nutritional traits in crops.

Nanomaterials Enable Delivery of Genetic Material without Transgene Integration in Mature Plants

Conference: International Conference on Plant Synthetic Biology and Bioengineering

Year: 2020

Proceeding: 4th International Conference on Plant Synthetic Biology, Bioengineering, and Biotechnology

Group: General Submissions

Session: Genome Scale Engineering

Time: Friday, October 30, 2020 - 12:50pm-1:15pm

Authors:

Landry, M., Chan Zuckerberg Biohub

Genetic engineering of plants is at the core of sustainability efforts, natural product synthesis, and agricultural crop engineering. The plant cell wall is a barrier that limits the ease and throughput with which exogenous biomolecules can be delivered to plants. Current delivery methods either suffer from host range limitations, low transformation efficiencies, or uncontrolled DNA integration into the host genome. Here, we demonstrate efficient diffusion-based biomolecule delivery into tissues and organs of intact plants of several species with a suite of pristine and chemically-functionalized high aspect ratio nanomaterials [1]. Efficient DNA delivery and strong protein expression without transgene integration is accomplished in mature *Nicotiana benthamiana*, arugula, wheat, and cotton leaves [2]. We demonstrate that transgene expression is transient and devoid of transgene integration into the plant host genome, of potential utility for easing regulatory oversight of transformed crops as genetically modified organisms (GMOs) [3]. We demonstrate that our platform can be applied for CRISPR-based genome editing for transient expression of Cas9 and gRNAs. We also demonstrate a second nanoparticle-based strategy in which small interfering RNA is delivered to mature *Nicotiana benthamiana* leaves and effectively silence a gene with 95% efficiency. We find that nanomaterials both facilitate biomolecule transport into plant cells, while also protecting polynucleotides such as RNA from nuclease degradation. DNA origami and nanostructures further enable siRNA delivery to plants at programmable nanostructure loci [4], which we use to elucidate force-independent transport phenomena of nanoparticles across the plant cell wall [5]. Our work provides a tool for species-independent and passive delivery of genetic material, without transgene integration, into plant cells for diverse plant biotechnology applications.

1. Demirer, G.S., *et al. Nature Protocols* (2019)
2. Demirer, G.S., *et al. Nature Nanotechnology* (2019)
3. Landry, M.P., Mitter, N.â€‹*Nature Nanotechnology* 2019
4. Zhang, H., *et al. Nature Protocols* (2020)
5. Zhang, H. *et al. PNAS* (2019)

Redesigning the Light Reactions of Photosynthesis

Conference: International Conference on Plant Synthetic Biology and Bioengineering

Year: 2020

Proceeding: 4th International Conference on Plant Synthetic Biology, Bioengineering, and Biotechnology

Group: General Submissions

Session: Keynote 2

Time: Saturday, October 31, 2020 - 10:00am-10:25am

Authors:

Hunter, N., University of Sheffield

Photosynthesis converts solar energy into chemical energy, underpinning the Earthâ€™s food chains and producing oxygen for respiratory electron flow. We can now envisage altering, enhancing and repurposing photosynthetic reactions, particularly in view of the need to support an ever-growing human population. This talk will focus on the light reactions, namely the absorption of solar energy and its transfer to the reaction centre where the energy is stored as a charge separation. A broad definition of redesign includes genetic engineering of light harvesting and reaction complexes for enhanced light absorption, re-routing pigment biosynthesis pathways to produce novel light absorbers, the de novo design of new light-absorbing and charge-separating proteins, and even the production of reconfigured photosystems immobilised on solid planar or electrode supports. The talk will summarise recent progress in these areas.

Characterization of Exogenous Mediators for Electron Extraction from in Vitro photosynthesis

Conference: International Conference on Plant Synthetic Biology and Bioengineering

Year: 2020

Proceeding: 4th International Conference on Plant Synthetic Biology, Bioengineering, and Biotechnology

Group: General Submissions

Session: Redesigning Photosynthesis

Time: Saturday, October 31, 2020 - 10:50am-11:15am

Authors:

Scherr, D., Virginia Tech
Senger, R. S., Virginia Tech

In vitro photosynthesis has garnered attention over the past decade as a novel system for generating an electrical current in a bio-solar cell. Our lab is focused on harnessing the reducing power of photosynthesis to catalyze in vitro enzymatic reactions, which requires a solid understanding of photosynthetic functions. Past characterizations of in vitro photosynthetic electron transport have relied on fluorescence measurements of chlorophyll pigments or electrical output of photosynthetic components [1,2]. Herein we describe a novel method for detecting reduction of exogenous mediators through ROS detection with the chemical luminol. Multiple quinones, whose structure and reduction potential vary, were used to characterize electron extraction from a photosynthetically active algal extract. The composition of the reaction buffer and the source of light were routinely adjusted to elucidate conditions that maximized electron harvesting. It was found that calcium greatly increased the reduction of the low redox potential mediator Anthraquinone 2-sulfonate (AQS), with other additives having varying effects on the different quinones tested. The light cycle used to stimulate in vitro photosynthesis and exposure to Triton X-100 also had profound effects on the reduction of exogenous quinones. Our findings can be utilized for tuning the environment of In vitro photosynthesis for the specific objective of a researcher, advancing the potential of bio-solar cells and in vitro enzymatic reactions as viable alternatives for energy.

1. Longatte, G., Rappaport, F., Wollman, F.-A., Guille-Collignon, M. & LemaÃ®tre, F. Mechanism and analyses for extracting photosynthetic electrons using exogenous quinones â€“ what makes a good extraction pathway? Photochemical & Photobiological Sciences 15, 969â€“979 (2016).

2. Hasan, K. et al. Photobioelectrocatalysis of Intact Chloroplasts for Solar Energy Conversion. ACS Catal. 7, 2257â€“2265 (2017).

Exploiting Underutilised Regions of the Solar Spectrum

Conference: International Conference on Plant Synthetic Biology and Bioengineering

Year: 2020

Proceeding: 4th International Conference on Plant Synthetic Biology, Bioengineering, and Biotechnology

Group: General Submissions

Session: Redesigning Photosynthesis

Time: Saturday, October 31, 2020 - 11:15am-11:40am

Authors:

Canniffe, D. P., University of Liverpool

Anoxygenic phototrophs use bacteriochlorophylls to harvest light outside of the visible range, relying on wavelengths unused by chlorophyll-containing oxygenic organisms typically found above in their ecological niche. Augmentation of oxygenic organisms with near-infrared absorbing complexes from anoxygenic phototrophic bacteria has been suggested as a means to increase the efficiency of photosynthesis in crops[1,2]. Current understanding of photosynthetic apparatus assembly and regulation, and the complexity of extending the chlorophyll pathway towards bacteriochlorophyll biosynthesis, are hindering progress towards this goal. I will present recent work detailing the successful transfer of a bacterial light-harvesting complex between purple bacteria, allowing the transgenic strain to access wavelengths that were previously unavailable. I will also discuss methods we have used for the modification of the (bacterio)chlorophyll biosynthesis pathway in phototrophic bacteria, yielding strains producing non-native pigments that permit photosynthesis in underutilised regions of the near-infrared. These approaches serve as first steps in the engineering of oxygenic organisms to capture light outside of the visible range.

1. Blankenship RE et al. 2011. Comparing photosynthetic and photovoltaic efficiencies and recognizing the potential for improvement. *Science* 332:805-809
2. Ort DR et al. 2015. Redesigning photosynthesis to sustainably meet global food and bioenergy demand. *PNAS* 112:8529-8536

Synthetic Carbon Dioxide Assimilation in Yeast for Autotrophic Growth

Conference: International Conference on Plant Synthetic Biology and Bioengineering

Year: 2020

Proceeding: 4th International Conference on Plant Synthetic Biology, Bioengineering, and Biotechnology

Group: General Submissions

Session: Redesigning Photosynthesis

Time: Saturday, October 31, 2020 - 11:40am-12:05pm

Authors:

Mattanovich, D., University of Natural Resources and Life Sciences

Global warming, caused by rapidly increasing atmospheric carbon dioxide levels, calls for novel measures to sequester CO_2. Assimilation of CO_2 into biomass and bulk chemicals is an attractive option. Besides the use of photosynthesizing organisms, synthetic biology offers the change to develop strains for a novel CO_2 based biotechnology. The methylotrophic yeast *Pichia pastoris* is widely used in the manufacture of industrial enzymes and pharmaceuticals. Like most biotechnological production hosts, *P. pastoris* is heterotrophic, growing on organic feedstocks. By addition of eight heterologous genes and deletion of three native genes, we engineered the peroxisomal methanol-assimilation pathway of *P. pastoris* into a CO_2 fixation pathway resembling the Calvin-Benson-Bassham (CBB) cycle, the predominant natural CO_2 fixation pathway. Due to the modular metabolic design which separates carbon assimilation from energy production, any NADH yielding energy source can be used. At present methanol oxidation is employed. The resulting strain can grow continuously with CO_2 as a sole carbon source at a Âμ_{max} of 0.008 h^{-1}. The specific growth rate was further improved to 0.018 h^{-1} by adaptive laboratory evolution. Mutations in evolved strains point at the balancing of CBB cycle reactions and NADH supply as critical steps for efficient growth. This engineered *P. pastoris* strain may promote sustainability by sequestering the greenhouse gas CO_2 and by avoiding consumption of an organic feedstock with alternative uses in food production.

Cr-Sphere: A Toolbox to Study Chlamydomonas Microbiota Interactions

Conference: International Conference on Plant Synthetic Biology and Bioengineering

Year: 2020

Proceeding: 4th International Conference on Plant Synthetic Biology, Bioengineering, and Biotechnology

Group: General Submissions

Session: Biotic Interactions and Synthetic Communities

Time: Saturday, October 31, 2020 - 12:00pm-12:25pm

Authors:

Flores-Uribe, J., Max Planck Institute for Plant Breeding Research
Duran, P., Max Planck Institute for Plant Breeding Research
Garrido-Oter, R., Max Planck Institute for Plant Breeding Research

Microalgae and bacteria interact in the environment by exchanging diverse nutrients and metabolites. Despite the relevance of these ecological interactions the association between soil microalgae and their microbiota remains understudied. To better understand these relationships, we use the model organism Chlamydomonas reinhardtii.

Co-cultivation of C. reinhardtii with soil samples revealed the recruitment of distinct bacterial communities by the microalgae. These bacterial communities overlap taxonomically with the core root microbiota of land plants, suggesting a set of conserved and possibly ancestral assembly principles. Furthermore, a culture collection of C. reinhardtii associated bacteria (designated Cr-Sphere) was established with representatives for 96% of the most abundant bacteria in the algal microbiota. Analysis of 200 genomes from these bacterial isolates revealed taxonomic and functional overlap between the Cr-sphere and other microbial culture collections isolated from plant roots. The Cr-Sphere can be used as a toolbox to study microalgae-bacteria interactions, their effect into the microalgal fitness, response to nutritional stress, or the establishment and evolution of microbial communities.

Biosynthesis of Cardiac Glycosides in Wallflowers (Erysimum, Brassicaceae)

Conference: International Conference on Plant Synthetic Biology and Bioengineering

Year: 2020

Proceeding: 4th International Conference on Plant Synthetic Biology, Bioengineering, and Biotechnology

Group: General Submissions

Session: Biotic Interactions and Synthetic Communities

Time: Saturday, October 31, 2020 - 12:25pm-12:50pm

Authors:

Jander, G., Boyce Thompson Institute

Accumulation of cardiac glycosides, small molecule inhibitors of essential Na^+/K^+ ATPases in animal cells, evolved independently as a defensive trait in at least twelve plant families. Although they are lethal at higher concentrations, plant-derived cardiac glycosides have a long history of use in both traditional and modern medicine as a treatment for congestive heart failure and other diseases. However, despite extensive research on the ecological functions and medical applications of cardiac glycosides, the complete biosynthesis pathway has not been identified in any plant species. Within the Brassicaceae family, the *Erysimum* genus is unique in its accumulation of cardiac glycosides. To provide new insight into cardiac glycoside biosynthesis, we sequenced the genome of *Erysimum cheiranthoides* (wormseed wallflower) and the foliar transcriptomes of 47 additional *Erysimum* species. Metabolite profiling detected almost 100 different cardiac glycosides, with up to fifty in some individual *Erysimum* species. Comparison of the cardiac glycoside content and a transcriptome-based *Erysimum* phylogeny provided evidence for rapid gain and loss of enzyme functions. There appears to be no tradeoff between the production of glucosinolates, characteristic defensive metabolites of plants in the Brassicaceae, and cardiac glycosides in *Erysimum*. We identified candidate genes for cardiac glycoside biosynthesis in *E. cheiranthoides* with a combination of ethylmethanesulfonate mutagenesis, co-expression analysis, and genetic mapping of natural variation in cardiac glycoside content. Both *Agrobacterium*-mediated transient expression and virus-induced gene expression silencing were used to confirm the *in vivo* function of the identified *E. cheiranthoides* genes.

Exploring Plant Alkaloid Biosynthetic Pathways in Yeast Using Untargeted Metabolomics and Proteomics

Conference: International Conference on Plant Synthetic Biology and Bioengineering

Year: 2020

Proceeding: 4th International Conference on Plant Synthetic Biology, Bioengineering, and Biotechnology

Group: General Submissions

Session: Engineering Bioproduction

Time: Sunday, November 1, 2020 - 10:30am-10:55am

Authors:

Jensen, M. K., Technical University of Denmark
Zhang, J., Technical University of Denmark
Kristensen, M., Technical University of Denmark
Schrübbers, L., Technical University of Denmark
Klitgaard, A., Evolva
Christensen, H. B., Technical University of Denmark
Keasling, J., Lawrence Berkeley National Laboratory
Hansen, L. G., Technical University of Denmark
Thieffry, A., Technical University of Denmark

Plants produce some of the most potent human therapeutics and other industrially relevant molecules. Despite the wealth of plant natural products with therapeutic effects, very few are commercially available because of low production levels in native plants, generally slow growth, as well as limited genetic tractability of plants that makes it difficult to engineer plants.

Reconstitution of biosynthetic pathways in a heterologous host is a proven strategy for rapid and inexpensive production of complex molecules. *Saccharomyces cerevisiae* is one of the most common choices for a heterologous microbial host because of the abundance of available genetic tools for these organisms. However, a deep understanding of both the heterologous natural product biosynthetic pathway and its connectedness with the native yeast metabolism is essential for developing cell factories with high catalytic flux towards natural product biosynthesis.

Monoterpenoid indole alkaloids (MIAs) are secondary plant metabolites that exhibit a remarkable structural diversity, with >2,000 MIAs derived from a common precursor. Many of these alkaloids exhibit pharmaceutically valuable biological activities, which makes large-scale production using recombinant microbial cell factories very attractive.

In this study, untargeted metabolomics and proteomics were used to investigate and compare the cellular metabolism and proteome of wild-type yeast and yeast engineered to produce MIAs. This has provided valuable insights into the changes associated with genetically modified biochemical reactions and metabolic pathways. Here we will present the learnings from our integrated omics approach using compound annotation to enable automatic selection and generation of target lists for LC-MS/MS analysis. We will also present the integration of LC-MS and LC-MS/MS data, and rapid compound annotation using a variety of different compound databases, as well as illustrate experimental designs and omics outputs leveraged for guiding novel MIA biosynthetic pathway designs and optimize bakerâ€™s yeast as a chassis for fermentation-based manufacturing of plant natural products.

A Modular Yeast Based GPCR Biosensor Platform for Compound Identification and Metabolic Engineering

Conference: International Conference on Plant Synthetic Biology and Bioengineering

Year: 2020

Proceeding: 4th International Conference on Plant Synthetic Biology, Bioengineering, and Biotechnology

Group: General Submissions

Session: Engineering Bioproduction

Time: Sunday, November 1, 2020 - 10:55am-11:20am

Authors:

Miettinen, K., University of Copenhagen
Kampranis, S., University of Copenhagen

G-protein coupled receptors (GPCRs) constitute the largest family of extracellular sensing entities in eukaryotes. The core GPCR mechanism is conserved across eukaryotes including the pheromone sensing pathway of bakerâ€™s yeast, *Saccharomyces cerevisiae*. We have developed a flexible modular biosensor platform by hijacking the yeast pheromone sensing pathway that in principle enables plugging in any GPCR. By optimizing a CB2 based cannabinoid biosensor we have achieved a sensitive system for cannabinoid detection with characteristics similar to those mammalian systems. As a proof-of-concept we used this biosensor to screen 71 Mexican medicinal plants for the presence of cannabinoids. Further fractionation of the plant extract and screening of the fractions enabled us to pinpoint the active compound(s) in the extract. Further expansion of this platform will be used enable the efficient optimization of cannabinoid producing cell factories.

Genetic Libraries of Cyanobacteria for Fundamental and Industrial Photosynthesis

Conference: International Conference on Plant Synthetic Biology and Bioengineering

Year: 2020

Proceeding: 4th International Conference on Plant Synthetic Biology, Bioengineering, and Biotechnology

Group: General Submissions

Session: Engineering Bioproduction

Time: Sunday, November 1, 2020 - 11:20am-11:45am

Authors:

Hudson, P., KTH-Royal Institute of Technology

Cyanobacteria are model photosynthetic microorganisms for studying light harvesting, electron transport, and the Calvin cycle. Cyanobacteria have also been pursued as potential â€œcell factoriesâ€□ for producing chemicals.

Engineering the Production of Medicinal Natural Products from Plant Tissue Cultures

Conference: International Conference on Plant Synthetic Biology and Bioengineering

Year: 2020

Proceeding: 4th International Conference on Plant Synthetic Biology, Bioengineering, and Biotechnology

Group: General Submissions

Session: Engineering Bioproduction

Time: Sunday, November 1, 2020 - 11:45am-12:10pm

Authors:

Lee-Parsons, C., Northeastern University

Plants have developed elaborate mechanisms for responding rapidly and specifically to changes in their environment. For instance, in response to attack by pathogens, insects, or herbivores, plants strategically redirect their metabolic resources from growth and development towards the production of defense-related specialized metabolites. This induction of specialized metabolism is also carefully controlled to limit the costly expenditure of resources. The overall vision of our research is to elucidate how plants rapidly and efficiently coordinate and control this defense response. Our model system for studying the regulation of defense-related specialized metabolism is the Madagascar periwinkle or the *Catharanthus roseus*. We chose *C. roseus* since it produces medicinally-valuable terpenoid indole alkaloids (TIAs) upon herbivory and deploys sophisticated mechanisms for the transcriptional regulation of TIA biosynthesis. TIA biosynthesis consists of ~30 enzymatic steps distributed across multiple cell types. The regulation of TIA biosynthesis is similarly complex as multiple transcription factors are required to coordinate the expression of multiple branches of the TIA pathway. This network is difficult to dissect since activators (like MYC2) not only induce the expression of the TIA pathway genes but also that of repressors (like JAZ). In the case of MYC2, its role in the regulatory network was obscured by JAZ until recently. Similarly, other relationships are likely masked by repressors since the overexpression of activators alone had a limited effect on TIA accumulation. In this talk, I will present the role of repressors in the regulatory network and how we can devise strategies to reprogram the regulation to enhance specialized metabolism.

Unlocking the Potential of Heunlocking the Potential of Hemicellulose-Derived Biomaterials

Conference: International Conference on Plant Synthetic Biology and Bioengineering

Year: 2020

Proceeding: 4th International Conference on Plant Synthetic Biology, Bioengineering, and Biotechnology

Group: General Submissions

Session: Engineering Form

Time: Sunday, November 1, 2020 - 10:30am-10:55am

Authors:

Urbanowicz, B., Complex Carbohydrate Research Center; University of Georgia

Polysaccharides are the most abundant biopolymers on Earth, largely as a result of the fact that they constitute both the major components of the plant cell walls that surround all plant and fungal cells, comprise the exoskeletons of insects and crustaceans, and play a major role in regulating the properties of the extracellular matrix in vertebrates. Replacing petroleum-derived materials with biodegradable plant-based polymers is a key strategy towards solving to this global problem. Currently, a large quantity of naturally biodegradable products are produced from starch, which competes with food production. At the same time, plant biomass derived from sustainable non-food crops contains polymers with great potential for use as plastics and other valuable materials. Hemicellulosic polysaccharides provide a diverse set of biocompatible building blocks for polymer science and engineering. Xylan is one of the most promising lignocellulosic polymers for use as a new feedstock for biodegradable products. At the same time, it is also one of the less studied natural polymers for applications that are relevant to the synthetic plastics and polymeric materials markets. This talk will cover some of the advances being made to develop chemical and enzymatic routes toward xylan valorization for potential applications in packaging materials, super-absorbents, and pharmaceuticals.

Decoding Development of Lateral Roots in Plants Using Recombinase-Based Logic Circuits

Conference: International Conference on Plant Synthetic Biology and Bioengineering

Year: 2020

Proceeding: 4th International Conference on Plant Synthetic Biology, Bioengineering, and Biotechnology

Group: General Submissions

Session: Engineering Form

Time: Sunday, November 1, 2020 - 10:55am-11:20am

Authors:

Guiziou, S., Centre de Biochimie Structurale
Nemhauser, J., University of Washington
Chu, J. C., University of Washington
Lanctot, A., University of Wasington
Gala, H. P., University of Washington

There are many open questions about the molecular mechanisms that coordinate the dynamic, multicellular behaviors required for organogenesis. By building synthetic circuits to either record or re-code in vivo signaling networks, synthetic biology brings potential novel inroads into studies of development. Serine integrases mediating site-specific and irreversible DNA recombination allow the implementation of history-dependent logic circuits. By developing an in vivo temporal tracker using recombinases, we will track and decode the cell-lineage gene-expression of signaling events involved in lateral root development in the reference plant Arabidopsis thaliana. To date, we have characterized a number of genetic parts to tune the level of integrase activity. In parallel, we are identifying promoters that are active during early lateral root development via single cell RNA-seq. The techniques developed here should serve as a proof-of-principle that can be extended to studying the development of other organs, as well as to developmental studies in diverse organisms.

Engineering Sustainable Chemicals Biomanufacturing Using Photosynthetic Bacteria

Conference: International Conference on Plant Synthetic Biology and Bioengineering

Year: 2020

Proceeding: 4th International Conference on Plant Synthetic Biology, Bioengineering, and Biotechnology

Group: General Submissions

Session: Sustainability

Time: Sunday, November 1, 2020 - 11:40am-12:05pm

Authors:

Jones, P. R., Imperial College London
Yunus, I. S., Imperial College London
Sattayawat, P., Imperial College London

The field of microbial photobiotechnology would benefit from success stories where genetically engineered alga and cyanobacteria are commercially used for the conversion of $CO_2$and N_2into non-native chemicals in a financially sustainable manner. At least for low value / large volume products such as fuels, photobiotechnological replacements may remain a dream as long as fossil fuels are subsidised and all direct and indirect costs are not paid for by the consumer. Photobiotechnology needs to supply large volume markets in order to have a measurable impact but the high-value sector also merits our current attention in order to get started and acquire commercial experience and confidence in the technology.

I will describe our group work on engineering cyanobacteria to produce a number of products, the lessons learned and the challenges still ahead. What could a roadmap towards commercialization of engineered cyanobacteria look like?

Selected references

1. Sattayawat, P., Yunus, I.S., Jones, P.R.* (2020) Bioderivatization as a concept for renewable production of chemicals that are toxic or poorly soluble in the liquid phase. **PNAS**117, 1404-1413.
2. Yunus, I.S., Palma, A., Trudeau, D.L., Tawfik, D.S., Jones, P.R.*(2020) Methanol-free biosynthesis of fatty acid methyl ester (FAME) in *Synechocystis* sp. PCC 6803. **Metabolic Engineering**57, 217-227.

Bioenergy Synthesis and Production from Microalgae

Conference: International Conference on Plant Synthetic Biology and Bioengineering

Year: 2020

Proceeding: 4th International Conference on Plant Synthetic Biology, Bioengineering, and Biotechnology

Group: General Submissions

Session: Sustainability

Time: Sunday, November 1, 2020 - 12:05pm-12:30pm

Authors:

Li-Beisson, Y., CEA

Microalgae are fast-growing microorganisms that have developed efficient mechanisms to harvest and transform solar energy into energy-rich molecules such as lipids. They are thus potentially great renewable cell factories for production of fuels and biomaterials for the chemical industries. However, several biological as well as engineering challenges need to be addressed before the establishment of a profitable microalgal industrial biotechnology. The main goal of our laboratory is to investigate the molecular mechanisms involved in the conversion of solar energy and atmospheric CO_2 to energy-dense compounds such as oil and hydrocarbon. We assess the molecular limitations of these mechanisms and propose *novel strategies* for synthetic biology of microalgae. The further use of this knowledge will aid in the design and creation of strains for algal biotechnology. In this communication, I will present current projects and highlight major findings from our group.

AUTHOR INDEX